Michael Roberts. August 74.

INVENTIONS

W. Heath Robinson

William Heath Robinson

INVENTIONS

W. HEATH ROBINSON

DUCKWORTH

First published in 1973 by
Gerald Duckworth & Co. Ltd.
The Old Piano Factory
43 Gloucester Crescent, London NW1.

Second impression 1973

ISBN 0 7156 0724 3

Printed in Great Britain
by Unwin Brothers Limited
The Gresham Press, Old Woking, Surrey, England
A member of the Staples Printing Group

CONTENTS

Domestic

Gastronomic

INTRODUCTION

It is time to present to the world a definitive collection of the comic fantasies of the great English artist William Heath Robinson (1872–1944), whose name passed into the English language during his lifetime. 'A Heath Robinson contraption' is a phrase that all Britons understand, even those too young to have any idea how the phrase originated; and this distinction is a rare one, especially rare perhaps with so extremely modest and retiring a man as Heath Robinson was. His vein of mechanical fantasy was consistent, original and precisely in tune with the fantasy of his age, his appetite for work was enormous, and the resulting body of artistic achievement is crystallised for all time, so long as machines remain machines and human beings need reminding that that is all they are.

William Heath Robinson was born in 1872, his father Thomas and his two elder brothers, Tom and Charles, all being professional artists. Between 1895 and 1915 he obtained many 'straight' commissions for book illustration, including *The Arabian Nights* (1899), *Don Quixote* (1920), *Hans Andersen* (1913) and *The Water Babies* (1915); but the first hint of the unorthodox and unique way in which his genius was to develop appeared in a children's book written and illustrated by himself—*The Adventures of Uncle Lubin*, published by Grant Richards in 1902, reissued by Chatto & Windus in 1925 and again by the Minerva Press in 1972.

Among the earliest admirers of his new fantastic vein was the Lamson Paragon Co of Toronto, whose London agent, Charles Ed. Potter, gave him his first commission for advertising drawings. Other early enthusiasts were Bruce Ingram, editor of *The Sphere*, and George Grossmith, who later got him to design a sketch for one of Charlot's revues; but success was by no means immediate. He was once rebuffed by an art editor with the wounding words: 'If this work is humorous, your serious work must be very serious indeed.' Fortunately the same drawings were received with laughter at the next place they were offered.

Another early admirer was H. G. Wells, who wrote to him on 31 December 1914: 'I have been ill and frightfully bored and the one thing I have wanted is a big album of your absurd beautiful drawings to turn over. You give me a peculiar pleasure of the mind like nothing else in the world.'

It was during the First World War that Heath Robinson went over almost entirely to comedy, collections of his anti-German cartoons being first published in book form by Duckworth in 1915 and 1916. 'The much-advertised frightfulness and efficiency of the German army,' he wrote, 'gave me one of the best opportunities I ever enjoyed.' His comedy was entirely without hate or malice, and he did as much as Bruce Bairnsfather and Bert Thomas—both of them friends of his—to sustain the morale of British troops at the front, from whom he began to receive endless letters suggesting new ideas and devices; many of these he adopted.

In 1918 he moved out of London with his family to Cranleigh in Surrey, and busied himself during the 1920s on behalf of a long succession of commercial firms which invited him, for advertising purposes, to illustrate their manufacturing processes, and elaborate them imaginatively in his characteristic manner. Coal mines, coke ovens, steel mills, Swiss Roll bakers, confectioners, papermakers, oil companies, asbestos cement, marmalade and lager beer manufacturers were among his employers, and he visited all their works to brief himself correctly. 'As the principles of mechanics are always the same,' he wrote laconically, 'this variety did not matter much.' He studied both their works and their workers, and was delighted with his warm reception everywhere by the latter. 'When their stern preoccupation with the work was lifted, these earnest men were like children out of school. Nothing pleased them more than to see that which had held them so tyrannically treated with levity.'

In 1929 he returned to live in London, in Highgate, and in 1934, now a famous man, was called on to design his own house—'The Gadgets'—for the Ideal Home Exhibition. In 1935 he produced *Railway Ribaldry* for the centenary of the Great Western Railway. The next year appeared the first of six small books on domestic subjects written with K. R. G. Browne of *Punch* and later with Cecil Hunt. In 1938 he made his first appearance on BBC Television, demonstrating a pea-splitting machine of his own design.

In 1939 the Second World War caught him at the height of his fame, and he harnessed his fantasy to its situations as readily as he had done in the First, the home front being this time as fruitful as the field of purely military operations. But as the war progressed he passed his threescore years and ten and the constant hard work, on top of his wartime volunteer duties and wartime privations, began to tell; he died on September 13th 1944 at home after an operation. He left a widow and five children.

If anything needs to be said about his comic art, it had better be in his own few words on the subject, quoted from his autobiography *My Line of Life*: 'Whatever success these drawings may have had was due not only to the fantastic machinery and devices, and to the absurd situations, but to the style in which they were drawn. This was designed to imply that the artist had complete belief in what he was drawing. He was seeing no joke in the matter, in fact he was part of the joke. For this purpose, a rather severe style was used, in which everything was laboriously and clearly defined. There could be no doubt, mystery, or mere suggestion about something in which you implicitly believed, and of this belief it was necessary to persuade the spectator. At the slightest hint that the artist was amused, the delicate fabric of humour would fade away. I do not pretend that this end was always achieved, but I was so far successful as to be frequently identified personally with my drawings. I was imagined by some people to be a kind of ingenious mad-hatter, wandering around absent-mindedly, with my pockets full of knotted string, nails and pegs of wood, ready to invent anything at a moment's notice. . .'

Almost the only other recorded remark of his about his own work bears witness to an underlying modesty not always characteristic of artists: 'It seems that an artist only plays a part in his creations. He focusses and passes on what was already in the universal heart of the race. Mankind and Michelangelo painted "The Last Judgment" '.

We might in passing note again the quality of innocence in his art, mockery without spite; that he has singularly few recurrent obsessions or mannerisms—cat burglars and bagpipes are perhaps two of them; that his ideas do not date—his ways of dealing with traffic problems and his recipes for community living are more relevant now than when he first drew them; and that each drawing requires detailed examination for some minutes if all its subtleties and refinements are to be appreciated. This was, in fact, what he demanded from his own wife and children, who were often summoned to the studio when a new piece had been finished, to tell him whether it 'worked' for them or not.

Perhaps the last word should be that of a fellow artist, J. F. Horrabin, who observed at their first meeting: 'He looked happy—like a man who has really found his *métier* and knows he has expressed himself in it.'

Such was the ingenious and engaging Englishman, whose work lives on to give us and succeeding generations its own uniquely flavoured delight in these pages.

M.H.

Acknowledgments

The above quotations from W. Heath Robinson: *My Line of Life* (1938) are included by courtesy of the publishers Blackie & Son Ltd. Further biographical material is to be found in Langston Day: *The Life & Art of W. Heath Robinson* (Herbert Joseph, 1947); *The Penguin Heath Robinson* (1966); and John Lewis: *Heath Robinson* (Constable, 1973).

The Pilsener Pump

Learning the Goose Step

American Barb Trowsers

Washing Day on Board a Zeppelin

The Boche Catcher

A Matter of Time

The Subzeppmarinellin

Hague Convention Defied—Button Magnets

Hague Convention Defied—The Tommy-Scalder

Hague Convention Defied—Use of Laughing Gas

Hague Convention Defied—Onion Whittling

The Lancing Wheel

The Trench Presser

A Trained Dog of War

Training a Carrier Pigeon

Bedroom Boat Race Practice

Cultivating Toughness in Footballs

Multi-Tennis

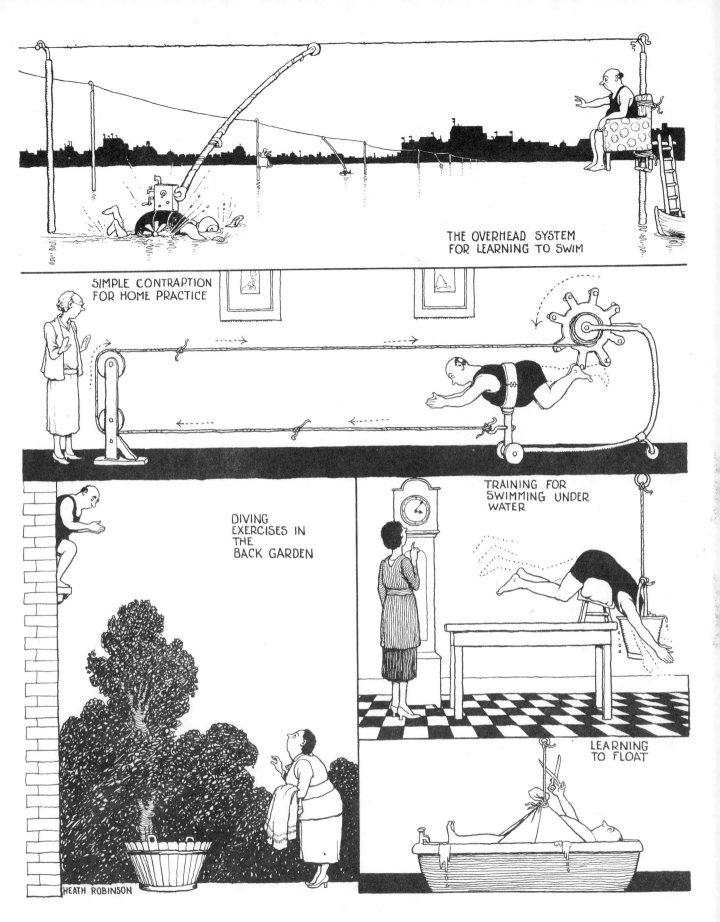

Learning to Swim

Ski Vagaries in Switzerland

Aero-Bathing Machine

Wimbledon Serving Tube

Safety First Cricket

Making Own Snow in Switzerland

Cricket for the Middle-Aged

The Aero-Widow

W. HEATH
ROBINSON

EXERCISE·1 *For accustoming the system to scrum pressure*

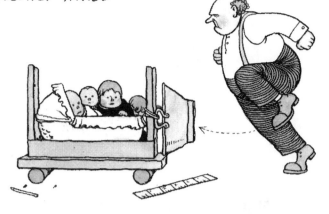

EX·3. *For inuring the body to blows from the boot*

EX. 6. *A problem for the more advanced·HOW to get out of a difficult position*

EX. 4. *For rendering the muscles of the neck pliable·*

EX.5. *For accustoming oneself to taps on the skull·*

Course of Training for Rugger Novices

27

THE "OUT" BELL
OR THE UMPIRE'S FRIEND

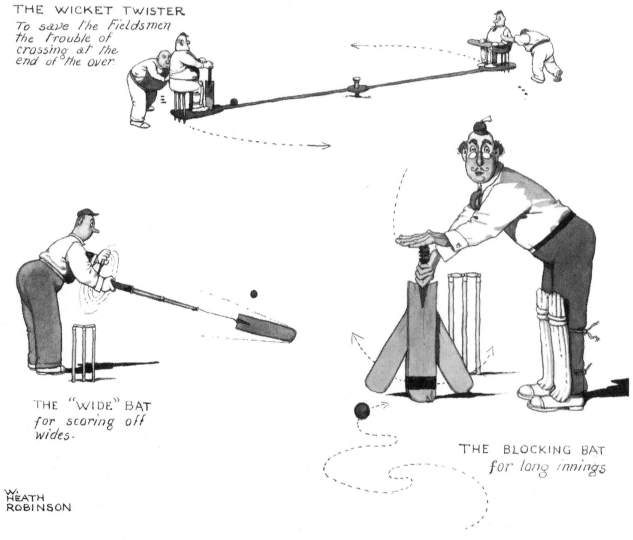

THE WICKET TWISTER
To save the Fieldsmen the trouble of crossing at the end of the over.

THE "WIDE" BAT
for scoring off wides.

THE BLOCKING BAT
for long innings

W. HEATH ROBINSON

Some Cricket Novelties

Training Polo Ponies

Self-Propelled Skating Gadget

New Season's Bathing Machines

Large Whist Party in Small Room

Foot Golf

Testing Golf Drivers

Trapping the Polecat

Bagging the Chipmunk

Decoying the Giraffe

Clam Spearing

Catching Young Tarpon

The Margate Fish Thief

The Habits of the Night Moth

Filming the Crocodile

Blasting Limpets on the Barbary Coast

Catching Foxes

Scent Discrimination in Foxhounds

New Combination Game

AN EXCITING GAME OF MIDGET FOOTBALL

A THRILL IN A GAME OF MIDGET BASE-BALL

A HAPPY DAY ON THE MIDGET
TENNIS COURTS AT WIMBLEDON

MIDGET POLO PROVIDES
EXCITEMENT WITHOUT
DANGER

Midget Games

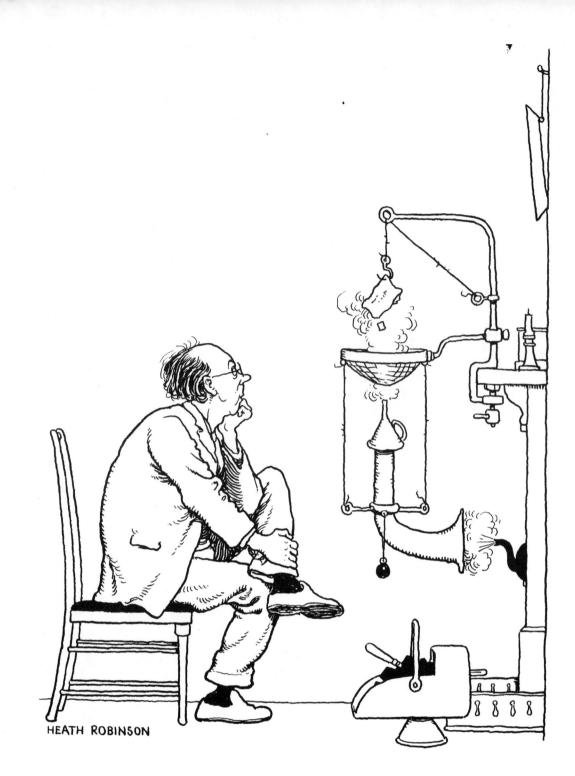

The Philatelist's Friend

Six-handed Draughts

The New Gambling Dance

Lighting a Pipe

The Aerocharibang

One at a Time Lock in Brompton Road

Portable Pedestrian Crossings

Candidates for Water Divining

Square Pegs into Round Holes

Anti-Litter Machine

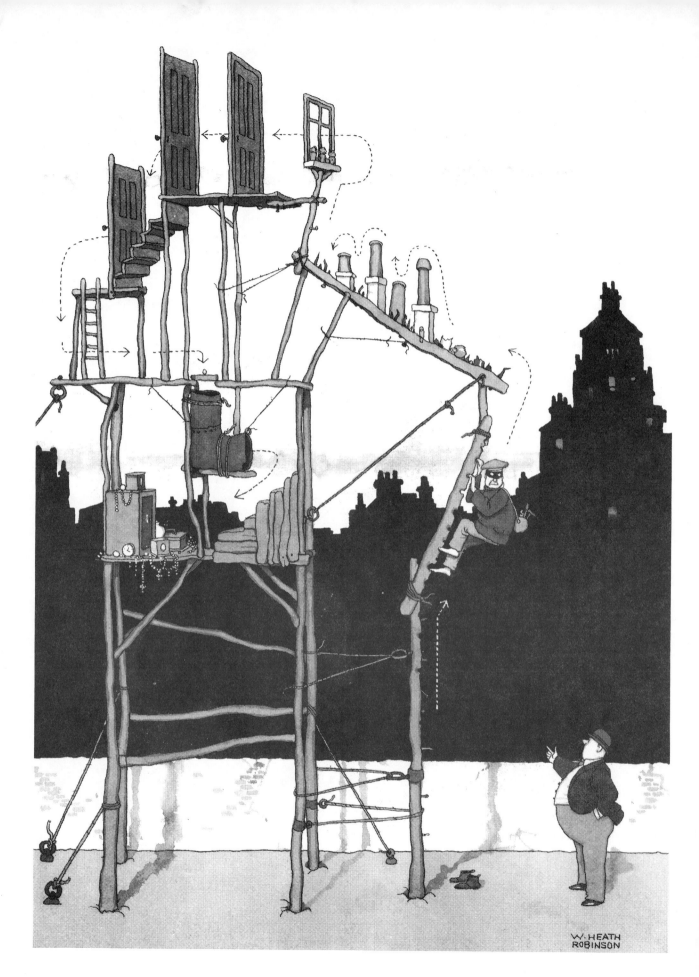

Training Frame for Cat Burglars

Testing Artificial Teeth

The Ascent of Mount Everest

Sidelights on the Wig Industry

Taking One's own Photo while Bathing

A Leak in the Channel Tunnel

Noah and the Flood

Air Dentistry

Aero-Larceny—Stealing a Dinner

Deceiving Nazi Dive-Bombers

Spies Stop Church Bells Ringing

Tobacco Rationing in Club

Camouflage on Salisbury Plain

An Incendiary in the Loft

Altering the Time on Big Ben

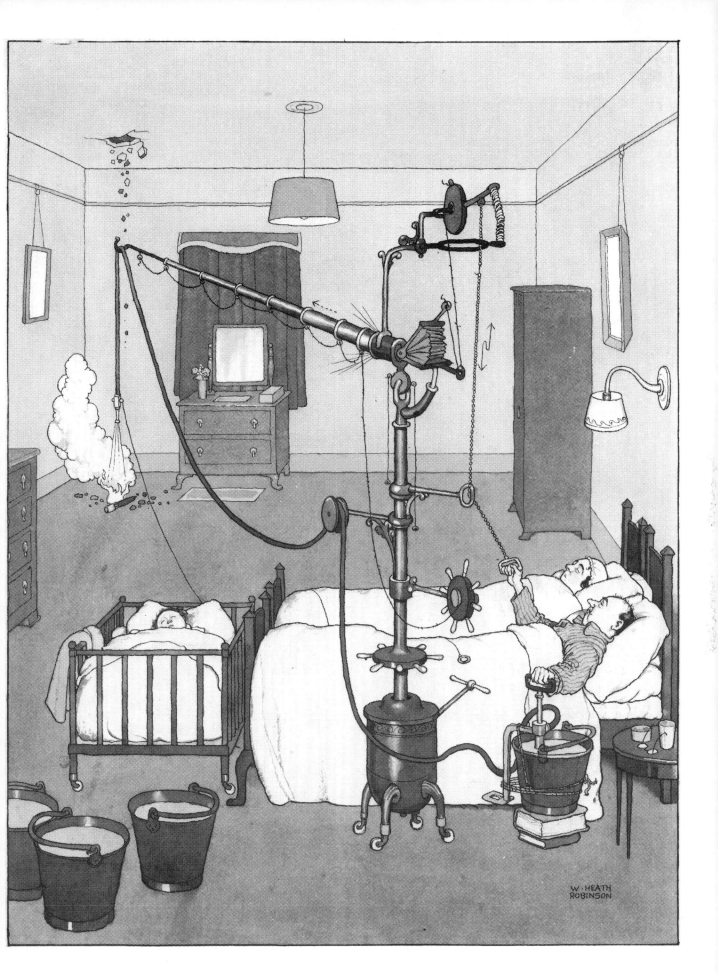

A Bedside Bomb Extinguisher

Confusing the Enemy's Sense of Direction

The Jerry Scarer

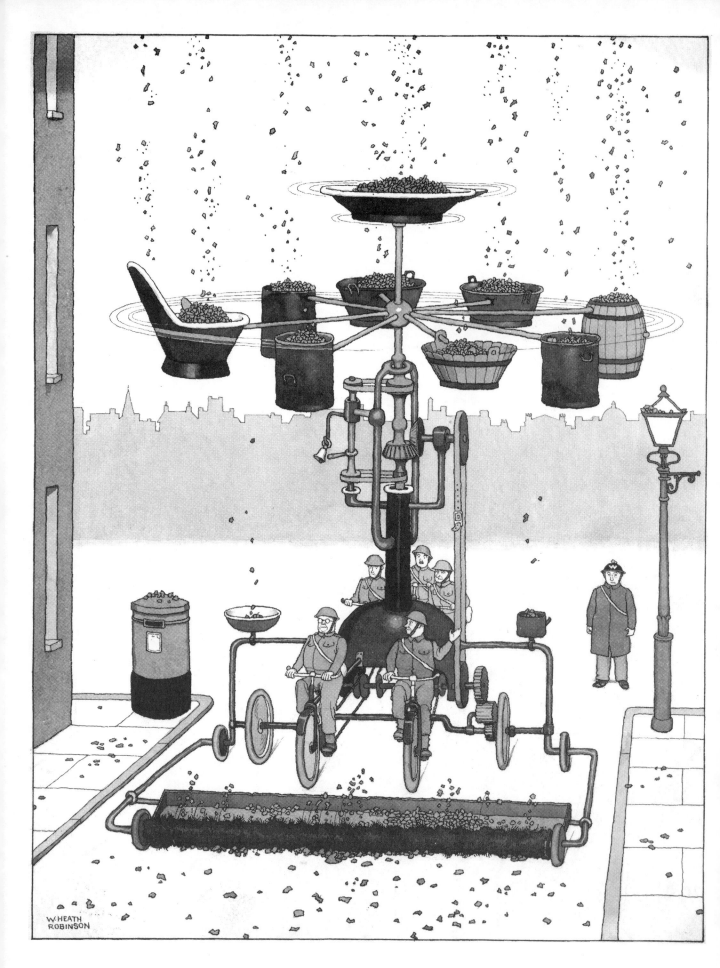

The Shrapnel Collector

Semi-Tanks

Triple Tanks

Back Acting Guns

Hit or Missler Gun

Delayed Action Bomb Machine

Realistic Invasion Exercises

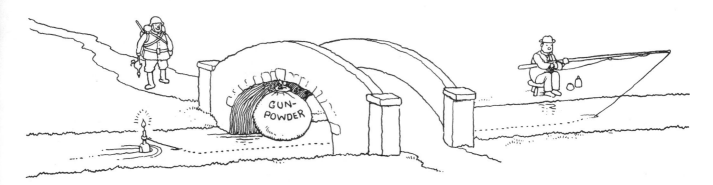

Primitive Guerilla Strategy

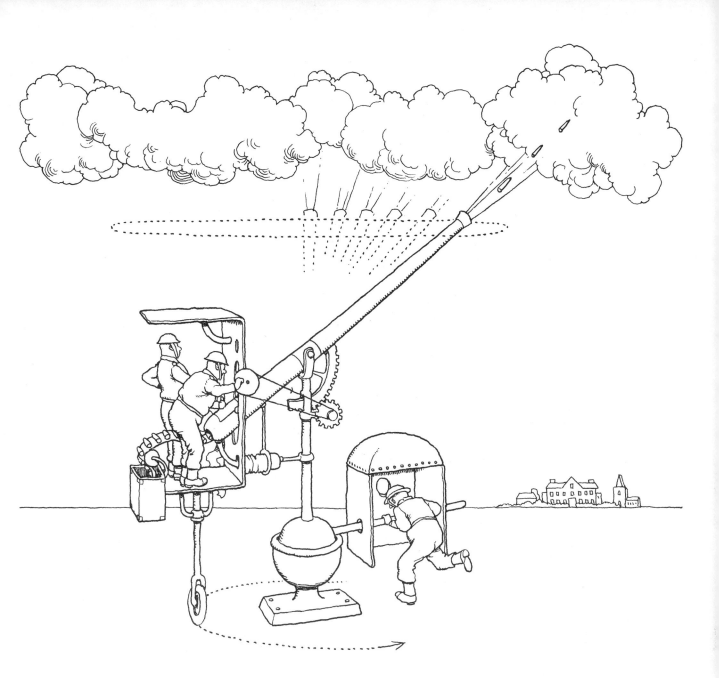

Ball and Socket AA Gun

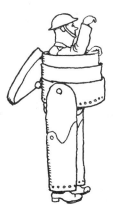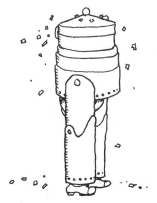

Shrapnel Trousers for Air Raid Wardens

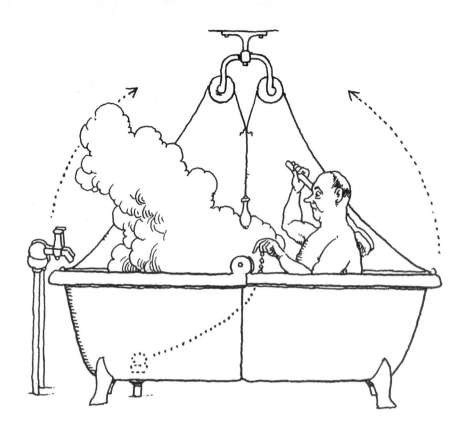

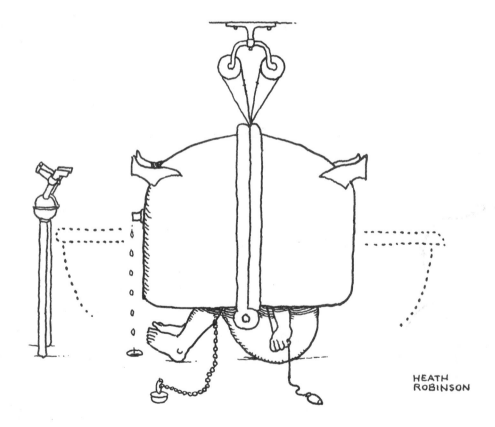

ARP Bath

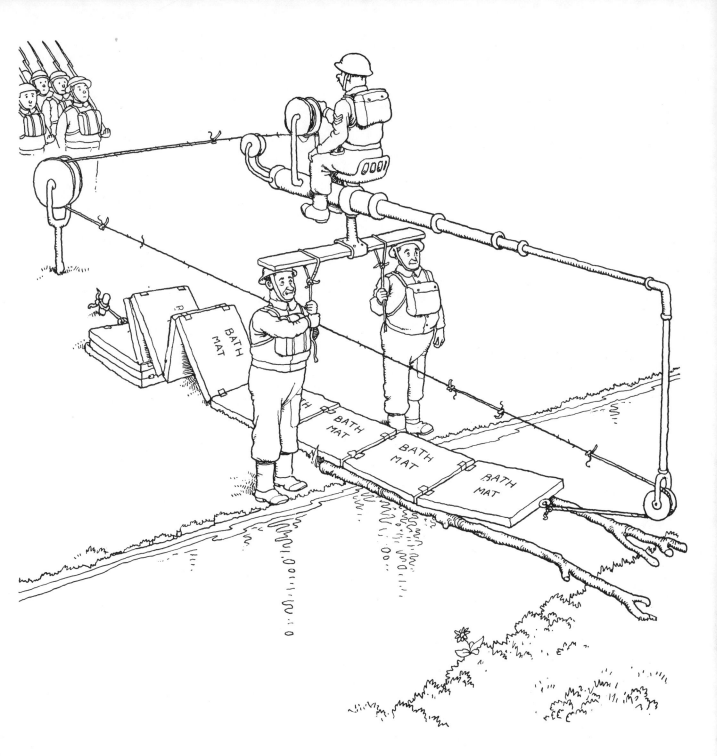

Cork Mat Method of Crossing Streams

Blowing Out Parachute Flares

Kidnapping Lord Haw-Haw

Anti-tank Trap Device

Luring a German Submarine to Dover

Arousing Special Constables

Camouflage v. Camouflage

The Trench-Sealer

The Camouflage Expert

Tank Skidders

Safe Practice for Paratroopers

Stale Ostrich Egg Throwers

Catching a Quisling

Cooling the Atlantic

The Jumping Tank

How to Cross the Rhine

The Naval Cloud Dispeller

The Incendiary Bomb Machine

The Multi-movement Bomb Catcher

The Macaroni Mine

Molotov Bread Basket

Tea and Sugar Rationing

Automatic Egg Rationer

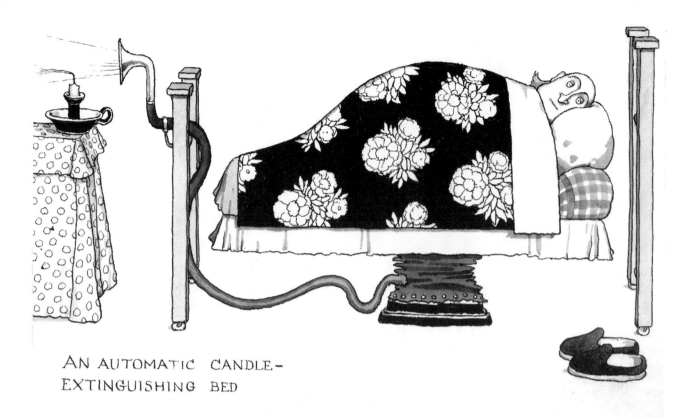

Automatic Candle-Extinguishing Bed

Top Watch-Winder

Capturing a Cook

Briquettes

Christmas Family Gliders

III

Orderly Conduct of Sales

Pram for Troublesome Children

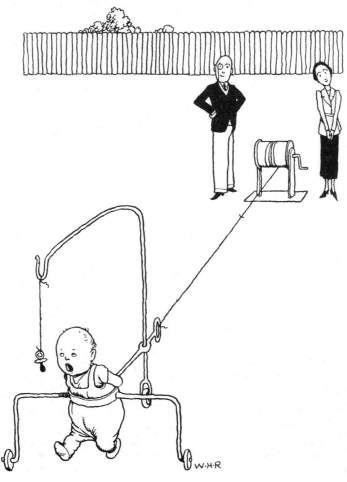

First Lessons in Walking

HEATH ROBINSON

The Fuse

Kitchen Chimney

No Soap in the Bathroom

Training a Child to Avoid Greed

Honey Wallah

Singeing Hair at Back of Head

The Perfect Husband Presses Own Trousers

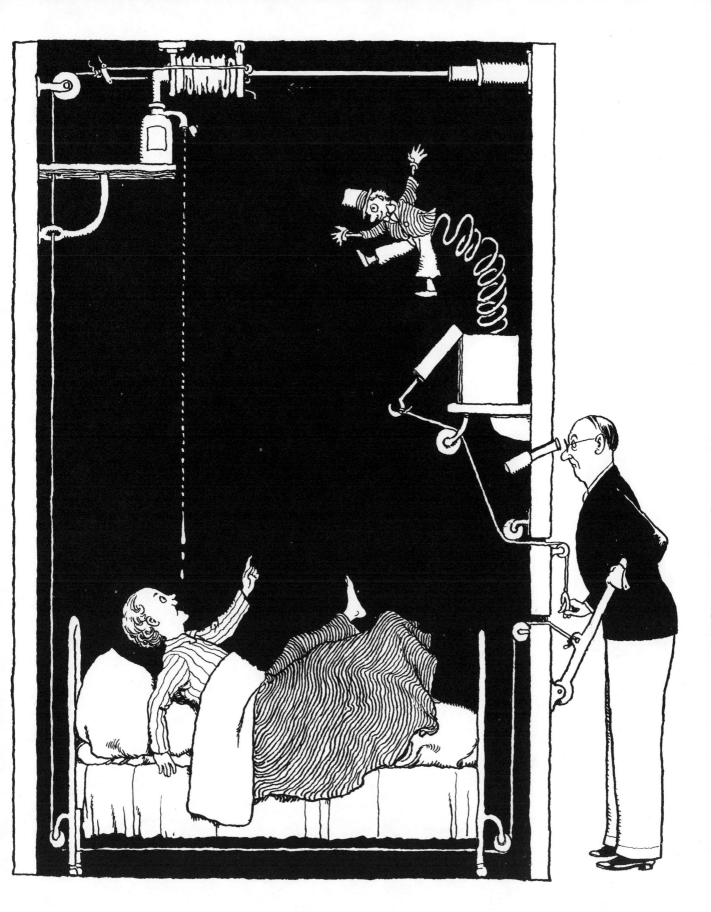

Administering Dose of Medicine to a Boy

Milkman on Early Morning Round

Disused Lighting—Heating—Cooking

Bedroom Suite de Luxe

Mass Piano Instruction in Communal Home

Breakfast in Bed for Housewife

The Soundness of Newton's Laws

Monthly Overhaul

More than One Thing at a Time

Six-tier Communal Cradle

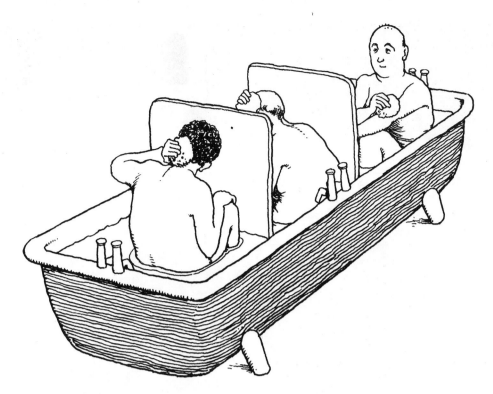

Three-man Communal Bath

Modesty Toilet Boxes

Community Bed de Luxe

Testing a Fiancée

Community Piano

How to Rise with the Sun

The Sunbathing Wheel

Jazz Bagpipes

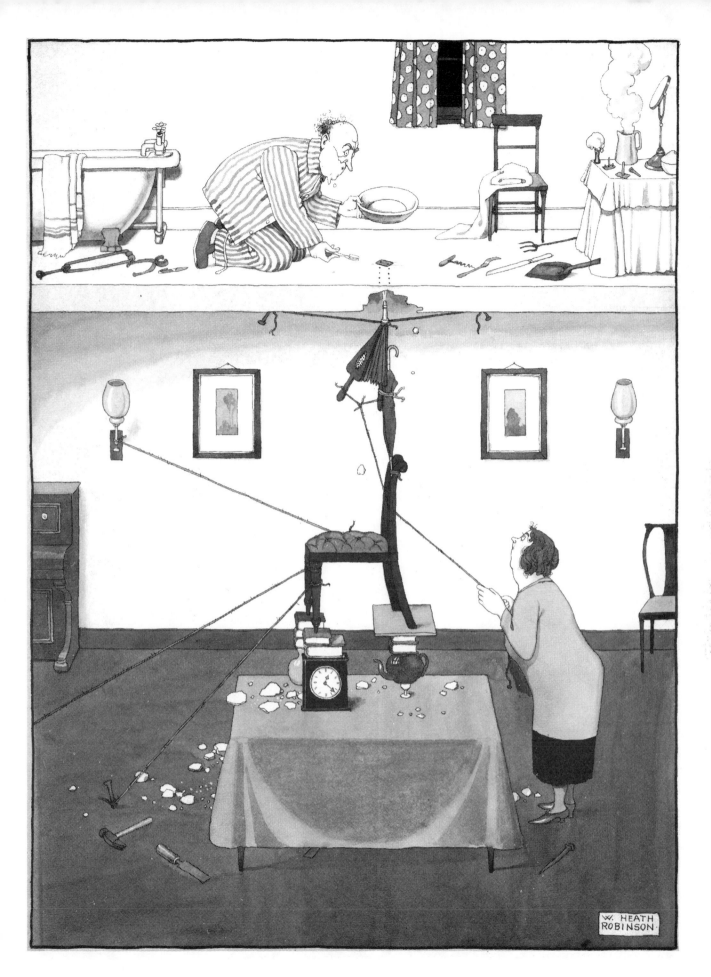

How to Pick up a Safety Razor Blade

Blowing out a Candle a Long Way from the Bed

Training the Bride's Train Bearers

A Communal Breakfast for the Young

A Surprise for the Cat Burglar

De Luxe Outfit for the Cat Burglar

How to Build a Bungalow

The First Spring Clean

Doubling Gloucester Cheeses

The Welsh Rarebit Machine

Stretching Spaghetti

Egg Armour Plating

The Champignon Bell

Busy Afternoon in the Whitebait Factory

Frittering a Banana by Electricity

How to Avoid Tears when Peeling Onions

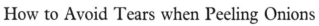

THE END